Alchemy
And The Tincture Of
Gold.

STEVEN SCHOOL.

ISBN: 1490957871
ISBN-13: 9781490957876

DEDICATION

I dedicate this book to the aspiring alchemists of our time. This art is earned by those who put forth the required effort, and those who quit are assured of only one thing, that they will never see the hidden light of our alchemical sun. In our time, we are a rare breed.

CONTENTS

ACKNOWLEDGMENTS

I would like to thank my friends in the alchemy world, your support over the years is greatly appreciated, and will never be forgotten. Over the hills we climbed, to see the sunlight through the rain. You know who you are.

1 THE BOOK OF THE SUN.

I would like to begin this book with an introduction to the alchemical definition of salt, sulfur, and mercury. It is very important to realize that the alchemists of old spoke philosophically, using metaphors. So the alchemical definitions of substances has nothing at all to do with the common, vulgar elements of the same name. in the ancient hermetic art of alchemy, anything which is liquid can be called mercury, since it flows like mercury, however elemental mercury is never used at all anywhere in our true art, those who experiment with real mercury are hopelessly lost in the labyrinth, they are far from ever finding the true methods, and most of them usually end up poisoning themselves, so stay far away from real mercury, and do not listen to alchemists who advise you in its use.

Now in alchemy there are two basic types of sulfur principles, to the alchemist anything which is a dry powder is considered to be alchemical sulfur, if this powder is combustible, then it is called combustible sulfur and this means that it is considered to be impure, because it cannot stand the fire of the fourth degree of heat which means bathed in flame, instead it has an impurity which adheres to it, that can only be removed by fire. Then there is sulfur incombustible, this is a dry powder which is alchemically pure because it is completely unaffected by fire. Our salt, is a volatile white substance, its appearance is similar to wax.

COUPLE OF QUICK REMINDERS, ELEMENTAL MERCURY IS NOT USED IN OUR ART, STAY AWAY FROM IT, ALSO DO NOT INGEST ANY SUBSTANCES WITHOUT A LICENSED DOCTORS ADVICE AND PRESCRIPTION.

There is a common term in alchemical circles which says salt, sulfur, and mercury, this has caused countless numbers of people to be led astray in their quest for the philosophers stone, since they mistakenly believe it to mean that there are three ingredients required to confect the stone of the mineral realm and because some persons mistakenly take the direct common substance instead of the correct philosophical ingredients.

The truth is that our aqua vitae, our secret water of life, our acetum acerrimum, contains both our philosophical sulfur, and our philosophical mercury, it is one liquid, that we will use to imbibe the salt. The sages often mixed up the names of these substances to confuse us, for example they sometimes called the salt; sulfur, mixing it up.

There are a few mathematical equations which have been attributed to the philosophers stone over the centuries, I will give you an in depth analyses of this which will make things in this area a little more clearer to you, in fact over the course of the book I will make these principles perfectly clear. There are in fact a few different ways to confect the mineral stone of the ancient sages, for the most part the differences between them is mostly the ingredients required, and the separation and purification methods utilized, my particular favorite method is called the poor man's way, the mathematical equation of this method is, from one is made three, and from the three, two are made into one, the third substance is not discarded, it is kept, for it contains great alchemical secrets within itself.

Now when you are reading the ancient alchemical writings of the sages, any time they make reference to quicksilver, it can mean either one of two things, neither one of which is elemental mercury. The alchemical meaning of quicksilver means the philosophers stone itself, the second alchemical meaning of this word is molten metal, when the sages said you must reduce metals to their first matter, this simply means to return the metal to its original molten state, so that it flows like quicksilver.

We seek to separate and purify two substances, the water and the salt, Sol and Luna.

Now the importance of this revelation is that to confect the stone, we do not combine philosophical salt sulfur and mercury, but instead by dripping the liquid onto the powder, a salt crystal will grow. The alchemist simply assists nature by separating and purifying the required elements, and then recombining them, and this is the second method of the process. The third portion of the work has been called conjunction, imbibition, and multiplication, in truth these three things are all one and it is a very simple procedure, modern science has a name for this portion of the work which will greatly simplify this part of the process, once you know what it is. In modern terms it is called growing crystals at home.

(The first portion of the great work we did in fact skip over here, however it has been sufficiently described in other booklets of this series therefore there is no need to reiterate that simple and primitive portion of the magnum opus here).

The substrate evaporation method of growing crystals at home.

This technique involves the use of a dry powder, and a saline aqueous solution. Once the correct substances have been alchemically prepared by art, the dry matter is placed into a round bottom flask, the liquid is then poured in, the substance is now stirred with a thin wooden stick until the dry powder is completely dissolved into the aqueous solution, now the top of the flask is sealed with a stopper The flask can then be placed in a mild sand bath at the temperature of warm sunlight, this degree of heat is that of a hen sitting on her eggs, which means body temperature, it is basically 98.6 degrees. You can also place the flask in a window sill where it will simply just absorb the rays of the sun, as the liquid is subjected to gentle coction, the substance will coagulate into a stone.

This is the method for confecting the red stone, the celestial ruby, which is the medicine for man, coagulated into a stone. The liquid, our aqua vitae, is dripped into a round bottom flask, just a little bit of the liquid, cover the top of the flask with a coffee filter and a rubber band to keep the dust out, and place this flask into the Egyptian fire. The liquid will evaporate into a salt which is clear, this beautiful salt will reflect a myriad of colors in the light and is the peacocks tail, it is also natures paint brush, as this salt is subject to gentle warmth, it will undergo color change, it is the coloring substance in the great work of the magnum opus, it has the power to tinge luna. However in this operation that we are discussing, once we have the dry salt, we will imbibe it with the same liquid that it was created from, and then seal the flask, place it back into the Egyptian fire and in time, the water will be turned into wine so to speak, because it will coagulate into the stone. Now this is the red stone for man. It can be multiplied by repeating the process.

The lunar phase of the great work, the white stone was confected by nature, thousands of years before we came along and began the great work, we do not need to create the lunar stone, we only need to free it from its prison, and purify it. Visit the interior of the earth, and by rectification find the occult stone.

Some alchemists sum up the great work in two phases which they call the lunar phase and the solar phase, and I have a few things to say about this, first of all this type of summary is incomplete, it has left out some of the process, and so far all persons that I have come across that used this terminology have turned out to be fakes, I am not meaning to say that all alchemists are fake who refer to the great work in this way, but let us look at the facts here, one so called modern "alchemist" recently told me that his work has been "stuck" in the solar phase for two years with no progress, I find this to be interesting because the solar phase of the magnum opus is nothing but a gentle evaporation/distillation of the matter being worked upon, if the workman has failed to complete this task in two years time, then there is something bad wrong with that operators understanding of our art, and that is putting it very nicely. The lunar phase comes after the solar phase, in the lunar phase we take our subject, and we distill it on higher heat, in this process we are preparing the white wife, Beya. This is a process of separation and purification which will prepare the female for conjunction with the male (Gabritius). The apparatus must be sealed for this portion of the work or we will lose our volatile spirit.

The solar phase of the great work is the process of preparing the red man, Gabritius, our living gold, sol, this is our water of life. it is now ready for conjunction with Beya, the androgynous matter, in the hermetical wedding.

A brief overview of the steps of the great work.

TRITURATION- to grind in the mortar and pestle.

EXTRACTION- the use of a menstruum.

EVAPORATION- collecting our extracted substance.

DISTILLATION- rectification, visit the interior of the extracted earth.

CONJUNCTION- the first imbibition, the hermetical wedding.

MULTIPLICATION- the repetition of the imbition/coagulation.

The great work could be summed up in three major groups or three main processes. The first portion of the work involves trituration, which means to grind our subject in the mortar and pestle, after which is performed an extraction, which includes a filtration and evaporation and this is the first portion of the work.

The second major portion of the magnum opus is the separation and purification of sol and luna through rectification. These substances must be kept in sealed flasks, it is the astral spirit and its earth. These substances are volatile, one more than the other, and they must also be kept pure.

In the third section of the work, we shall perform the conjunction, adding the aurum solis to the white stone, in order to tinge luna, and coagulate it into the citrine stone.

There are some other minor or secondary operations which are performed and they have already been discussed in my other books, but here we are focusing on the main operations which is the bulk of the art.

A few questions have been asked of me lately by aspiring alchemists from across the globe, I may as well include the answers from those questions here, in case anyone else is seeking them. As concerns the prima materia of the ancient sages, it is a certain middle substance which exists in nature, over time it changes form and evolves, therefore it must be collected at the proper time in order to be of any use to us in our art, for further clarification on this matter, study my booklet, alchemy and the green lion, gently contemplate that entire booklet, one section at a time, until you understand this issue. Alchemy and the green lion is the starting point in this series of booklets on the magnum opus, I have seen people going through periods of frustration while diligently studying that work, and I have literally watched that frustration transform into illumination as they began to understand the truth of the words written therein;. Which opened the door to a deeper understanding of nature's mysteries. These moments of illumination are truly divine revelation, as each new level of wisdom is earned by the workman.

Now another question recently arose, wherein a person who studies my booklets, asked another "alchemist" some questions about our art, this person stated that he had never successfully confected the lapis philosophorum, yet he also said that all methods publicly listed for free all over the internet, all are correct paths to the stone, first of all, his methods all resulted in failure so obviously all the methods are not correct, and second of all, is a person who failed to confect the stone, qualified to teach you how to confect the stone which they themselves could not do?

There are also a few people out there spreading false recipes which are toxic, I have said it adamantly in my other books and I will say it to you again,

STAY AWAY FROM TOXIC SUBSTANCES! AND STAY AWAY FROM THE "ALCHEMISTS", WHO ADVISE YOU TO USE THAT JUNK.

Now, another bit of advice about becoming hopelessly lost in the "labyrinth", if you are going to study alchemy from a teacher, choose one that has been successful in the great work, because with an instructor who failed, the best you can hope for is that they will bring you up to their level, which is FAILURE!

Ask to see pictures of some of their laboratory work as proof that they are qualified to instruct you, since a picture is worth a thousand words. Furthermore, stick with one instructor only, since if you ask three different persons a question, you more than likely will get three different answers and then you will become hopelessly confused, especially if the persons providing the answers happen to be a mix of people some successful, and some unsuccessful in our art. Therefore if you are following my work and you have a question about my work, then simply refer to my other booklets for the answers, they are a series. Most of my books are numbered, it is best to start with alchemy and the green lion, contemplate that work until you achieve the illumination concerning the secret contained within that writing, and then move on to alchemy and the golden water, etc. it is best to keep my booklets in good condition, because in times of limited availability the value of these works tends to fluctuate dramatically. At one time I had taken alchemy and the golden water out of print, I then saw a lady in Germany selling her copy of it for close to $350 bucks, in mint condition. When I am gone, my writings will no longer be available and the existing copies will be worth their weight in gold.

I was also recently asked what equipment do I use for metallic experiments since that particular alchemist did not want to purchase an expensive furnace if it was not needed, I myself have been down that road, I no longer possess a furnace for smelting metals, I have an assortment of miniature cast iron cauldrons with lids, as well as an assortment of miniature cast iron frying pans, and a small, portable plumbers torch which is commonly used for soldering copper pipe, but which also works conveniently well for melting the soft base metals. I have crucibles, crucible tongs, and extra long, thick, leather gauntlet type welding gloves for carefully handling hot objects.

Let me also give you a little advice on the writings of the sages, their writings were not designed to teach you what you seek, they do not show the way, or provide a proper understanding of alchemy, the writings of the sages were specifically designed to keep you lost in the labyrinth forever. Some people say it nicely when they tell you that the sages spoke philosophically or that they wrote in metaphors, well I have news, a metaphor and an outright lie looks a lot alike in the realm of the ancient sages. There are also plenty of modern "alchemists" that seem to continue the tradition.

In pursuing the great work, it is imperative to have great patience, mastering this art is not something which is done quickly, in fact one could spend an entire lifetime and never run out of experiments to perform, or new levels to explore in the creation of matter. Do not expect to learn in three weeks what took me five years of hard work to attain, however I have expedited the process for you by providing my insights into the matter, so it should move along much faster for you, than it did for me.

An interesting note that I would like to mention, I have seen miniature cast iron cauldrons for sale online that have designs on them, one of them has a circle with a crescent on each side of it, in witchcraft circles this symbol is commonly called the triple moon, I have a bit of news for you, it is not a triple moon symbol at all, instead it is an ancient recipe or formula if you will, for the confection of the philosophers stone, now the symbol does not show the processes leading up to conjunction, but it does show the proper ingredients, as well as the proper amounts to be combined. The symbol is one sun and two moons, it means one part of sol is combined with two parts of luna. The salt and the water, the salt represents our philosophical silver which will dissolve like ice in warm water which we discussed earlier.

The Chinese yin and yang symbol also represents the alchemical hermaphrodite, sol and luna. Our rebis, or two thing.

Something I read a long time ago comes to mind right now at this very instant. As you enter the city you will be greeted by a man carrying a pail of water, follow him to the house that he enters, the true meaning of this statement has just dawned on me, another illumination or divine revelation, because learning this art with no teacher was very difficult for me, it took a tremendous amount of effort, but here this simple statement clearly points the way from the beginning.

There is one portion of the writings of the sages that speaks the truth as clear as a bell, and that is their pictures, everything was color coded to them, even their family crests, these things contained the family alchemical secrets as plain as day for those who understand, just as an electrician has color codes which he uses for his work, so does the alchemist. I once saw a family crest from previous centuries in which the shield was color coded black, white, yellow, red. This is the color code of the great work, now there are some secondary colors and we could list each and every one of them in order to be more confusing, but it is more honest and upfront to list the four primary colors and they are in the correct order.

A recent break through that I have made in this art is called mercury of ten eagles, in many alchemy circles people tend to think that each "eagle" means one distillation, some even believe that mercury is taken at face value which is a gross misunderstanding. First of all, elemental mercury is never used in our art, so stay far away from any and all toxic substances, our subject is simply a crystal clear aqueous solution, the ten "eagles" is not always ten distillations, but instead it can mean ten parts to one part, now an eagle has wings which means that it is a volatile substance, so it can mean ten distillations, but it can also mean ten parts of a particular substance

now earlier we discussed that the correct mixture is two parts of luna to one part of sol, and this is also very true, for each imbibition is roughly two parts to one part, however we still have to multiply the stone, which means that the imbibitions must be repeated. So if I apply the aqueous solution to the salt, and complete one turn of the wheel, then each time I repeat this will be one multiplication, and since each additional imbibition is only a matter of adding additional liquid, and not more of salt, then each repeat turn of the alchemical wheel will bring us closer to our ratio of ten to one. This ratio is also a rough guide, it does not have to be exactly perfect, but it does need to be fairly close, anywhere from nine to twelve parts of the liquid to roughly one part of the salt will suffice, and there are two ways that we can perform the imbibitions, we can give the subject all of its water at once, or we can use the method of repeating the smaller imbibitions, as the stone grows larger it will require a larger portion of the liquid for each imbibition.

Sendivogius liked to apply all of the liquid at once which takes longer, Flamel and Paracelsus advised in their works to use the repetitious method. The imbibition and multiplication process can very easily become confusing when you begin to approach that phase of the work for the first time, however it quickly becomes child's play with the knowledge that I am about to share with you, the stone itself is a salt crystal, which is grown at home, using simple methods and we will discuss some of that in this book, if you would like to gain a better understanding of it than this book offers, you can simply do an internet search on "how to grow crystals at home", this little bit of knowledge will greatly speed your ascension through the levels of our art.

A good portion of our work is simply separating and purifying our elements, and then the process of wetting the salt with the aqueous solution, when the matter reaches dryness, it is wetted again.

I was recently asked about the effects of aurum solis, I will say that the medicine does not heal injuries, if you suffer from physical issues, it cannot heal that, but old aches and pains seem to miraculously disappear, and injuries do seem to heal faster, but if you fall off a building, it is not going to help.

STEVEN SCHOOL

"mercury" of ten eagles has been and can be done in two different ways, for example, Paracelsus and flamel each had a different method and I will describe both of them, flamels method is top notch, Paracelsus method is much faster. First Theophrastus method, he separated the substances and purified them, (Sol and Luna), just as Flamel did, and then Paracelsus went straight to conjunction.

In flamels method, he calcined the earth leftover from the green lion to redness, put the liquid back onto this powder, and distilled it again, this was repeated ten times, and was known as the method of the ancients which flamel learned from the book of Abraham eleazer. So Flamel performed these ten distillations, and then proceeded to the conjunction.

When I began my study into the hermetic arts, I spent about four years contemplating the green lion. I wondered is it under the vegetable roots as some of the philosophical writings of the sages say?, is it brought to earth in springtime through dew and rain?, is it the secret substance which makes the grass green and causes the vegetables to grow?

I remember driving down the highway one fall day out in the country, I was looking out the window at the trees and grassy meadows as I was wondering what this green lion of the alchemists could really be, this entrance to the philosophers garden, this secret substance which is called chaos. As my journey brought me into the city I passed a road side billboard which read Jesus, victory over chaos.

This ancient stone, this celestial ruby, all of my life my heart has told me that this stone is not a myth. I have seen pictures of religious figures such as Jesus and Mary holding a heart over their chest with a flame coming out of it, the eternal flame, the flame of eternal life. Years ago I had no clue what this symbol meant.

It is the secret fire, the celestial water of life, our aqua vitae, our philosophical, alchemical "mercury". NOT ELEMENTAL MERCURY!!!

I WILL REMIND YOU TIME AND TIME AGAIN, STAY AWAY FROM ALL TOXIC SUBSTANCES. IF YOU STRAY FROM THAT ADVICE, THEN TAKE RESPONSIBILITY FOR YOUR OWN ACTIONS.

Over the centuries many of the sages including Paracelsus as well as others, have left us clues in their writings.

Paracelsus himself said "let fire and azoth suffice for you". In another writing he basically repeats this statement but changes the phrase as well as the names of the substances, "it is not the common mercury and sulfur which are the matter of the metals".

Herein he is giving us another piece of the alchemical puzzle, so what he means by this is that the "fire" is also called "sulfur", which means that it is a dry powder type of substance, and that "azoth" is also compared to "philosophical mercury", which means that it is a liquid, and also that it is not toxic elemental mercury, but instead a crystal clear liquid which some have called dew.

This means that we require two things, our living gold and our living silver, Sol and Luna, Gabritius and Beya, the red man, and the white wife.

From the description of these substances it is very plain to see that one is a dry earth, and therefore must represent the body of our philosophical stone and that the other substance is a liquid, which therefore must be the multiplier, and is used to perform the imbibition's, and is nature's paint brush, which she uses to color all things, the trees, plants, rocks, metals, crystals, gemstone, even thee and me.

A brief reminder in case I forgot to mention it earlier, the free recipes for confecting the stone which are posted all over the internet, are all false, every single one of them, I have also seen one "alchemist" who has placed a new twist on these free recipes by mixing and matching them, which means that he or she, takes an ingredient from one recipe, and a vessel from another, and then takes a portion of a process listed in another "free" recipe, then combines all of this together, and charges $30 to "teach", this great work to you, while admitting that there has never been any success achieved through that particular technique. The work itself in that case, leads to a complete and utter failure, and most persons are smart enough to recognize it for what it is.

HERMES TRISMEGISTUS.

Our founding father, our father of lights, the father of alchemy.
It has been said that you are the master of the three worlds, alchemy, magic, and astronomy.

But I say unto you sir, that although I believe this to be true, I see it in a different light. Where others accepted what was written for the multitude and in their acceptance they accepted defeat, like the Trojan horse of the city of troy.

But sir, I say to you, my opinion as an alchemist is that as master of the three worlds is to be understood the plant kingdom, the animal kingdom, and the mineral kingdom.

As over the centuries so much lies and disinformation has been spread far and wide that in my time, since I prefer to write a book of truth, where others only spoke lies and deceit, I would like to therefore simplify the mineral kingdom by simply calling it the metallic kingdom or the metallic realm. We do know sir that it still covers the crystals, the gemstones, and the metals, but we wish to eliminate confusion and so we will evolve since we must know the true and principle differences between these three kingdoms if we are to proceed and advance in our art, which many are eager, willing, and deserving, among the multitudes of those who do not deserve our hidden and sacred knowledge.

Our father who art in heaven, hallowed be thy name, but for our Mr Trismegistus, who worshipped Amun Ra, the god of the sun, interesting it is that Jesus is the supernatural son of the sun, it is as if Jesus has been here many times, and we were a little to slow to learn, so he has great patience in his love for us, which Paracelsus summed up by saying that, "you will transmute nothing, until you have first transmuted yourself". By which he simply means that you must have the utmost patience and understanding, if you are to succeed in our art. So let us now proceed, the names may mingle but enlightenment is always our goal.

For Jesus so loved the world, that he set chisel in hand and cast his words in stone, for those who have eyes to see, for all time.

That sulfur, is the seed of gold, is true.

That our sulfur is no ordinary sulfur is also true.

That over the years, many persons have mistakenly chosen iron sulfate as the green lion, is also, another fact.

No one comes to the stone, but through the green lion, and no one approaches the red lion but through the green. If you would know the hidden light of our alchemical sun, you must know these things.

Sol and Luna, yin and yang, positive and negative, our living gold, and our living silver, the light and the darkness, victory over chaos, the secret and sacred knowledge, of our living god.

The green lion, is the entrance to the philosophers garden, this is also the garden of Eden, sol and Luna, represent the tree of life, and the tree of knowledge.

This sacred and secret garden, is protected by a flaming sword which points in all directions. You will not touch it, or gain entrance here, without god has seen into the stony earth of your heart, and granted permission, to this sacred and spiritual place, where our mercuries and tinctures abound.

As Mr Jesus Christ himself said, "my kingdom is not of this world".

He simply means that, the garden of Eden, is a sacred and spiritual place, where man may not physically walk, but only in spirit, and only then, if he is pure of heart.

There is something that I want to touch on while it is still somewhat fresh in my mind. A few pointers. Many of the sages have spoken of two substances being the key ingredients of the stone. In Sendivogius new chemical light, he mentions that the second substance is the multiplier and not the first, he also says that concerning the two substances, one congeals in the cold, and the other congeals in heat. All of these little tidbits have telling facts in them, these little pointers becoming quite telling when you get to that level of the work. In the writings of Gualdus I once found a description of him confecting the stone, it said that he was grinding sulfur in a mortar and pestle, as a yellow oil was dripping over the helm from his retort. This is rather telling about which substance was produced first, and which was produced second, it also is revealing as to which one used heat to be collected, and that also reveals to us, which one must have been congealed in the cold. Now I dare say that when he mentioned grinding sulfur in the mortar and pestle, he did not mean common sulfur, which does have its place in our art, but not at this particular juncture. So we know here that the oil is the second substance, and Sendivogius has clarified for us that this then is the multiplier. In the writings of Paracelsus he says that this oil can be yellow or red, it is actually capable of changing color which is caused by heat. Further in his writings he says to place this oil onto its salt by drops, then bake it in the Athanor for twenty days and it will be congealed into a stone, now Paracelsus is said to have pioneered shorter and much faster methods of confecting the stone than his predecessors, and so it appears that his method of multiplication takes twenty days to complete each turn of the alchemical wheel. I have also noticed that the writings of some of the sages seem to contain a small portion of the work, almost like someone wrote out the entire process of the magnum opus, and then separated it into individual paragraphs and then inserted them individually into other books at random, so that they are hidden, and when they are found, they really don't make any sense unless you find several of them. I have found pieces of this puzzle that I have described in the writings of Paracelsus, Michael Sendivogius, Alexander Seton, Nicholas Flamel, George Ripley, and Gualdus. Each one has hidden in their writings at least one paragraph, that matches the other, picking up where one author left off, this is very, very, interesting to me, it most definitely does suggest something deeper, some type of link between these persons, other than just the fact that they were alchemists.

now I myself do not practice the urine work in alchemy, but I have researched and experimented with it in the past. The philosophers stone is not found in that substance, in truth, the ancients used it to create a very subtle, yet powerful menstruum which burns as fire, and was their secret key to unlocking the metals. It does in fact work exceptionally well for extractions. Here is how it is made, urine is left in glass jars for a few months, they are sealed with the seal of Hermes, which means that it is not fully a perfect seal and can actually breath just a little bit. After putrefaction, gently distill this vile substance per retort in sand bath on low heat, place a rubber stopper on the distillation arm of the retort, and affix a receptacle to this, this is a long, slow, gentle distillation of about three days. The clear distillate comes over the helm and is stored in a sealed glass, next the rest of the liquid and phlegm in the retort is taken outside and poured into a calcination dish, the stench is horrid, so it must be done outside. Using an electric hot plate, evaporate this to dryness on med/low heat, do not boil it over the sides of your vessel. Once dryness is achieved turn up the heat and calcine it very well. Let it cool, grind it in a mortar and pestle, then return it to a calcining dish, bake it out doors until it issues forth no more smoke or stench from the ashes, now let it cool.

Now extract the centric salt from these calcined ashes, you can use distilled water, or white vinegar. Place it in a jar covered with saran wrap, shake the jar a little each day to stir the matter within the glass. Let it sit for about a week to perform a really good extraction, now filter the liquid, and evaporate it to dryness to expose the incombustible salt. At dryness, calcine it again, then repeat the process of dissolve, extract, filter, evaporate, calcine, the point is to get the salt pure and clean.

Once this goal is reached, rejoin the clear distillate to the white salt in a clean retort with its attached receptacle, gently distill forth the clear distillate on low heat, and then return the distillate back onto the salt residue in the retort, now repeat the distillation process, this shall be done ten times, which will result in a subtle, fiery spirit, which is a very good menstruum, powerful and penetrating. This can be used to extract an essence from a substance, and then this very volatile liquid can easily be evaporated away which will leave you only with the extracted essence, clean and pure. The sages say that less eagles can be performed if you are working with the lesser metals, and that this menstruum of seven eagles has power to sway the moon, (silver), and that of ten eagles has power to calcine the sun, (gold).

I once tried this myself, I used the distillate of one eagle, and placed it into a sealed flask with a penny, over the course of a few months of time, this liquid extracted a tincture from the penny, without damaging the penny, the water had become a beautiful azul blue color, I suspect that it had extracted copper sulfate which is toxic, but the interesting discovery that I had made was that the menstruum does in fact work very well, and can be applied to other substances.

If you are going to attempt this method, use small placer gold nuggets from the river, pound them flat and thin, then place them into your menstruum of ten eagles to digest for several months until the liquid is the color of gold, now take some of your purified white salt of urine, place it into a round bottom flask, drip the golden colored liquid onto this salt by drops, let it evaporate while keeping a filter over the flask to prevent contamination from dust. Repeat these Imbibitions in order to completely saturate the salt with the tincture of gold. The liquid menstruum, will gently open the pores of the metal, extracting the metallic seed, then the goal is to store this in a suitable incombustible salt, and evaporate the menstruum.

This method does in fact take some time to complete, however it is simple and straight forward and is in fact much simpler than the poor man's method. Now in other books I have stated that gold is not needed to make the stone, and that is very true, but for this particular recipe it is needed, and there is in fact more than one way to the stone, so while other recipes do not require or use gold and silver, this is basically a reverse process of getting to the stone. One method begins with the basic elements and builds them up to the finished metal, the other way begins at the finished metal and works backwards so to speak. However, each method has its benefits.

I will also mention, that the process which nature uses to make gold in the earth, utilizes substances being digested by bacteria, to reach the final stage of gold, this represents putrefaction, this stage has already been completed by nature, so when you get to conjunction, with this particular method, there will be no black stage. Do not expect it or be waiting for it because it will never come as nature already completed it long before you were even born.

Most of my booklets focus upon the poor man's path to the stone, and this section of this work is focusing upon the rich man's method, these are two very different paths, so if one statement contradicts another, take that into consideration, that one statement may either refer to a different path, or to a different process.

Now, we have been discussing the rich man's method of confecting the stone, and I have also mentioned that most of my work involves the method known as the poor man's way, which is aptly titled because it does not require common gold.

Here I would like to discuss and clarify some of this work, we have discussed cohobation in some of my previous works, it basically means to pour the liquid onto the solid substance, it can mean the same then as conjunction, however there is a difference between the two, and this is a very important piece of the alchemical puzzle.

As I have said all along, alchemy can be very confusing so I try to keep it as simple as possible, I myself at times still tend to become a little befuddled in my words when discussing some of the aspects of our art. These substances that we are working with have been called by hundreds of names, the names have also at times been mismatched and reversed, calling sol luna, and luna sol, the sages did this on purpose to lead everyone astray, now with that being said, there are two lions, one green and one red. During our process we will come across three substances, here they are in order, the water, the white powder, and the red powder. The red powder is then placed back into the retort, its own water is added back to it, it is redistilled, and this is repeated ten times. The ancients had a name for this which I do not like to use because it is misleading, so let us just call this process, "oil of ten eagles". Now the process that was used to obtain the white powder, notice we have not here divulged that secret, but that portion of the process is listed in some of my other alchemy booklets which are a series, this is repeated a second time upon the white powder, so that it is as pure as the driven snow. Back to the water and the red powder, by the time the tenth eagle is completed, our substance has now attained the color of a translucent ruby, and has achieved a distinctive fragrant smell.

During the reading of alchemical tracts which are widely available throughout the internet and other resources, it can very easily become quite confusing for a myriad of reasons. First of all, not only did the sages call single substances by hundreds of different names, but they also mixed up the order of processes, they listed secondary colors, instead of sticking to the four primary colors of black, white, yellow, and red. Now we know that there is also both a dry path and a wet path to the stone. So if you are reading about a process written hundreds of years ago, more than likely it does not bother to mention whether it is part of the dry path, or the wet!, this single fact alone has kept many potential alchemists from progressing in this art. References to the threefold magical fire, the yellow oil, or the red oil, are a very strong indicator that the process being alluded to, is a portion of the dry path. The wet path has several additional steps and processes which must be completed, and this is why the dry path is in fact much faster than the wet, however the wet path is superior to that of the dry, and is the generally preferred method of many alchemical adepts.

The Flamel Path. Nicholas Flamel was a great alchemist, some people even venture to believe that he faked his own death, that his mausoleum may have actually been empty, and that he may even still be around today.

 Irregardless of these issues, many persons in alchemical circles often speak of the flamel path, which we will be covering later on in this book, so I would like to take this opportunity to explain something, in order to avoid confusion when we get to that particular process.

Any alchemist who reached a level in his work of being able to complete the Magnum Opus, is conversant and understanding in both the dry path, and the wet, since they are but opposite sides of the same coin, to fully and completely understand one, is to fully and completely understand the other. Many operators therefore had experience in both methods, and when an alchemist reaches adept level and finds themselves at this juncture, many of them choose the wet path, for the sheer beauty and gracefulness of it, as it flows smoothly from one process to the next. It is plainly obvious to me that Mr Flamel had experience in both methods. Some of his writings clearly indicate to any adept that he was following the dry path, yet some of his pictures also show vessels used for the wet path. So one person will tell you that Nicholas practiced the dry path, and another will tell you that in fact he used the wet method, both are correct.

Paracelsus was also very fond of the wet path. In some of his writings he alludes to having shortened the work, he liked to tease with his metaphors and unfinished statements. Often alluding to something great, and then leaving the reader to wonder. Well the fact of the matter is that I have been studying the writings of many of the sages for many years now, along with my work, and I have discovered some great secrets. We will at some point in this book discuss and compare the flamel and Paracelsus wet methods, and easily clarify exactly how Mr Paracelsus shortened the work, when he indicated that persons would have obtained their goal much faster, if they had studied in his school.

Another double meaning that I have frequently seen in the mineral works is mercury of one, two, three, four, five, six, seven, eight, nine, or ten eagles. Now our mercury can mean several different substances in alchemy, none of which are common mercury, or cinnabar, which are never needed or utilized in our work. In alchemy, mercury is simply just a liquid, however it is also sometimes used to describe a solid substance. Generally any reference to mercury of any number of eagles can mean either one of two things, it can mean for instance, ten parts of this, to one part of that, or it can mean a certain number of repeated distillations. Now the ancients did in fact use urine in their work, and this has created a huge stumbling block for many aspiring alchemists who believed that this then, must in fact be the veritable prima materia itself, which is simply not the case at all. Picture basic plant spagyrics, in which you have a substance, and a menstruum, the agent and the patient, in regards to the stone of the mineral realm, let me give you a direct quote from the sages themselves, found deeply hidden in the writings of some of the masters of this art, "the menstruum is distilled urine". So if you are searching for the philosophers stone in urine, you will never find it, because you have only the menstruum. Now Hermes himself is said to have advised, that mercury of seven eagles has power to sway the moon, and mercury of ten eagles can calcine the sun, otherwise you're menstruum will not be worth a fig. this quote is not in exact words, but we will get the point across, the old ones putrefied urine, and then gently distilled off the clear distillate per retort at roughly body temp, after this was completed, the matter remaining in the retort was calcined outdoors, (baked), until there was no more smoke, fumes, or stench emanating from it, it was then dissolved in distilled water, which was sometimes collected as rain water in pots, the water containing the dissolved ashes was then filtered and evaporated to dryness, this dry salt was placed back into the retort, the clear distillate was reintroduced, the matter was then redistilled at roughly body temp. this process was repeated and each repetition was known as the flight of one eagle, or mercury of one eagle, (distilled urine is meant, not elemental mercury), with each additional flight, the menstruum becomes stronger, until it becomes as a subtile spirit which burns as fire, and has the power to dissolve and extract. Another quality of this particular menstruum, is that it is so volatile!, it completely distills over the helm at body temperature, so therefore it must be stored in a sealed flask until ready for use, once this menstruum is used and has completed its mission to dissolve and extract from the chosen substance that it was applied to, it is then filtered, and gently evaporated, leaving you with only your pure extracted essence. The menstruum itself has completely separated itself from your substance, and then vanished into the air, its task complete. Flight of ten eagles, will produce a great menstruum, which can be used upon many things, however flight of one eagle, and your menstruum, will not be worth a fig.

I realize that I may seem a little repetitious in this booklet, perhaps in time my writing style will change, who knows, but this work is meant for alchemists, and the secrets are here, this is not a romance novel or a mystery thriller. It is simply, an overview of the great work.

Let us now, delve a little further into the hermetic mystery, let us digest some food for thought. There are two paths two the stone that I am concerned with here, irregardless of whether or not the possibility of other methods exist, these two I am familiar with as all true adepts are and this is what we shall work upon at this juncture, the dry path, and the wet. Now the wet path uses requires two things at its starting point, the substance, and the menstruum, the menstruum is the agent, which acts upon the patient, it is the liquid menstruum, that makes this the wet path, and which performs the actual work upon the substance, so I ask you then, how could we change this equation to cause this to be a dry work, right here at the beginning?, there is only one way correct?, therefore if the beginning of the wet path requires these two substances, then the only way to change this to the dry path right here and now, would be to eliminate the menstruum, and delete it from this equation. Which can only mean that as the wet path first begins with two substances, the dry path begins with only one substance, and since we are at the very beginning of the work, the substance in question at this time, could only be, the veritable prima materia itself, the rough stone. We know that as a preparatory measure in the great art we generally begin by grinding in the mortar and pestle, once our substance is prepared to begin, (ground to dust), how will we instigate change in the matter, how will we cause a reaction to occur if the liquid menstruum has been deleted from this experiment?, and once we answer this simple question, what then will be the output from this action?, what reaction will be generated from it?, once we contemplate and answer this simple question, it becomes very obvious what it is that we need next, and after this, we are then left with what the alchemists called the threefold magical fire, the great salt mountain. Interesting, three substances. Yet if we reach this point through the wet path, at this particular juncture, we will be left with only two substances. This explains why some of the sages spoke of mercury and sulfur, and some of them spoke of salt sulfur and mercury, suffice to say that this simply means that some were doing the dry path, and some were doing the wet path, even so, in the old hermetical drawings, sometimes is seen, three rivers coming from one root, then two of the rivers rejoin while the third flows off on its own, the dry path and the wet path are simply two different methods of preparing, separating, and purifying the matter, preparing sol and luna for conjunction in the hermetical wedding. So whether you chose to follow the dry path, or whether you took the wet path, we still arrive at the same junction, which is the chymical wedding. We now are left with only two purified and exalted substances. Although the wet half of this equation will appear accordingly at this juncture, as it was made. Since there were two paths to bring us here, the certain variables involved brought our substance to different levels, the real only difference is only in one of the two substances, and the difference is color, and I am sure that by now, we all understand what it is exactly, that affects color, in the great work. We chose which stroke of natures paintbrush that we desired, before we collected our substance at the beginning of the work.

If you are not following along here, if you have somehow lost track, and are not getting the message, then please, by all means, take the time to pause and reflect all that I have said, and remember, that truth is simple. Even though I walk through the valley of the shadow of death, I shall fear no evil, for my rod and my staff they comfort me.

There are three directions that we can go with the two substances Sol and Luna, this is indicated in old alchemical pictures, the number three is found quite often. What I mean by this is Aurum Solis, the white stone, and the citrine stone. These three things, are the treasures of the alchemists.

PROVERBS 3:16

Blessed is he who finds wisdom, she is more precious than jewels and nothing you desire compares with her, long life is in her right hand, in her left hand are riches and honor, all her ways are pleasant, and all her paths are peace.

Two substances are required in the great work, both of which are volatile, this is why in the statement vitriol, which roughly translates to visit the interior of the earth, and there by rectification find the occult lapis, rectification simply means purification by distillation, volatile means that a substance can evaporate, which makes it a very good candidate for purification by distillation. Here lies the big difference between the dry path, and the wet path. We know now that the process of separation and purification of sol and luna, involves distillation, the wet path requires a menstruum, so this portion of the work is the same as basic plant spagyrics, however in the dry path, no menstruum is used, this portion of the work is eliminated. And that is the true difference between the dry path and the wet. My booklets are like a set of study aids, they simply point out some alchemical facts, as I discover them, to help point the way, they may help you if you are also on this path of seeking the stone.

 it has been brought to my attention that there are some contradictions in my books, after this book is published there definitely will be a few contradictions between this writing and some of my previous works, so let me explain something to you right here and now. To err is human, the learning process is a system of trial and error, we experiment, we make a mistake, we learn from it, we move on and advance. To err is human, it is not really a wrong doing, it is simply a characteristic of being mortal. As I advance to new levels in the great science I write new books sharing some of what I have learned, so this book is written at a much higher level of understanding than when I wrote alchemy and the green lion for example, which is also a good book. But realize that when one finds the stone, this is not the end of the work, but only the beginning, once you find the stone, only then can you begin to experiment with it and delve into its mystery, as I have said many times before, there is no magic tooth fairy that shows up and explains the medicine for man or the medicine for metals when you find the stone, so at this juncture you must continue to study and experiment, delving into the higher levels, I only found the stone because early on in my search I quickly deduced that the current "teachers" of that time were pathetic frauds and fakes, I learned to recognize that during the course of their works, they had in fact discovered a few interesting facts, but then intermingled it with star struck fantasy land garbage, so the frustration of being confronted with bouts of ignorance from the so called adepts of the time period, drove me to break off on my own, and to conduct my own search, my own study, without being indoctrinated with someone else's misconceptions and mistakes. Now if a teacher learns a mistake, then they will teach the mistake to the student, and if the student looks up to the teacher, with admiration and great respect, then they words of the teacher are as good as gold to the student, and the mistake being taught, is readily accepted as iron clad fact, set in stone, but this particular point right here, stops all further progress for the student, as they are now thrown off track, and hopelessly lost in the labyrinth, which creates a huge obstacle, impeding further progress.

So, we were discussing the two substances, sol and luna, our philosophical gold and living silver, the water and the salt. Now some things that I currently understand much more fully, than in previous writings, I am going to share with you now, we are dealing with two substances here, a liquid, and a white salt, the liquid is sol, and is aurum solis, the white salt is luna, Paracelsus said in his writings, let fire and azoth suffice for you, by fire he means this volatile white salt, by azoth he means the volatile liquid. These things were not easy to figure out, because the writings of the sages were designed to confuse, not to teach. Paracelsus and his friends were simply taunting us, if they had wanted to, they could have explained things in much simpler terms. We cook our "fire", in our "water."

Now my books, guides, and booklets, are simply study aids, they are designed to assist you and point you in the right direction in your own search for the stone, some have given up in the search and blamed me for their own mistakes, others have persevered and put forth the required effort and earned the stone for themselves, I have made it much easier for you to find the stone on your own with the writing of these books, for in my search I had no teacher, and figuring out the correct substances and the correct processes was the most difficult task that I have ever undertaken in my entire life, I did not have to write this for you, I could have left the pages blank and then the world would have been left in darkness for all time, but it was I who found the stone, amidst all of todays "adepts", it was I who brought this knowledge back into the light, after it had been lost for countless centuries. My books are like a home study course, enabling you to follow along with my work, sharing with you the processes and paths which I took, that brought success in this art, where countless thousands of others had failed. Now with that being said, some persons have disputed the price of my writings, it is like two sides of the coin, is knowledge worthless, or is it priceless, if I were to teach you the stone in a classroom, what price would you put on that?, enrolling in college, tuition fees, even if I taught it online, it would have to be done in a series of lessons, each one with a price tag. My book commissions range from 20 cents at the lowest up to $5.62 at the current highest, and everywhere in between. So being that each of my writings, is a home study lesson, in a highly specific field, of a very rare and lost art, and being that I am qualified to teach because I achieved success in this art (which has been repeatedly proven), where so many others failed, what is it really worth for each individual lesson that I teach in the mineral realm of the great work?, so when you buy one of my books, booklets, or guides, you might pay the purchase price of the book, plus shipping and handling, (unless it is a kindle edition), and you might pay any applicable sales tax, let us now theorize and hypothesize that perhaps this total amount could conceivably escalate to a total of twenty dollars or less per book, let us now ponder, what is it really worth to have a direct lesson, from one who has actually found the stone that we seek, is it worthless?, or is it priceless?, or is it somewhere in between?, is there a happy medium?, I myself have studied very hard, for several years, I have conducted hundreds of experiments, during the early days of learning the work, I must have spent at least ten grand on laboratory apparatuses, books, and substances to experiment upon, it was both a difficult work, and an expensive study, but I found the stone, I reached my goal, I am very happy with these results, that I earned for myself, through diligence and hard work. My books tend to be short because they are not filled with useless fluff and filler, they are a series because out of the goodness of my heart I decided to share my findings with you, but at the same time my animals deserve to be fed, they would like some food on their plate, I am happy to teach, but I prefer my animals to be very well cared for. Mr Flamel was a bookseller, even though he also knew great secrets, I too would like to be a bookseller, is it really that hard to fathom?, in her right hand is long life, in her left hand are riches and honor, is it really that hard to understand, that I choose, the right hand, and thus have no care for the left, as long as my animals are well fed? How much do I really need?, the humble book seller suits me well. To each his or her own, you may choose whichever hand you like, or both even for that matter, I would like to follow more closely with Mr Flamels example, except that I teach and share knowledge much more openly.

So to reiterate, we have two paths, the dry and the wet. These are the processes of the preparation of sol and luna, the separation and purification, of the liquid and the salt, from their impurity. A two part distillation, beginning in the retort on low heat, and ending in the aludel on higher heat. A two part rectification, beginning with distillation, and ending with sublimation. The spirit and the body, the water and the salt. Sol and luna, our philosophical gold and our living silver, the liquid is our aqua vitae, the water of life, the medicine for man. The salt is the white stone, if the two are combined and subjected to gentle coction (cooking), then together they become the citrine stone that is the medicine for metals, the body and the spirit, the aqua vitae is the real holy water. I have confected three stones, I confected the red stone from the liquid alone, this is the medicine for man, before conjunction of sun and moon, the dose is very small, it is only one grain, as in the size of a grain of salt, it is dissolved in liquid, it is allowed to set for a minimum of four hours, and then it is filtered, this liquid must be diluted to the color of liquid gold, with no redness, it is taken only twice per year on the vernal equinoxes, unless one encounters a disease or sickness, the normal dose for a long and healthy life, is two drops of this liquid in a beverage, taken twice per year on the vernal equinoxes, in the case of disease or serious illness it is said to be up to one tea spoonful of the properly filtered and diluted medicine per day until the sickness is cured, know this to be a rare secret, because if the medicine is not prepared according to these guidelines, then either it will be too weak, or it will be too strong and fire the body. In this example of "fire" the body, the sulfur principle, luna, the white salt, is what Paracelsus means when he says therefore let fire and azoth suffice for you, therefore the claims of false adepts who said the white stone was medicinal, are incorrect, it is only sol, the liquid. I remember reading long ago, too long ago for exact quotes, but basically what was meant, they who not knowing, ingested thinking that they had found the medicine, but with it they took death instead. This is a very good example, do not ingest any tinctures until you are sure of what you have, and what are its possibilities. For many have said that the white stone heals issues of the mind, and the red stone heals issues of the heart, restores youth, causes a strengthening of the limbs, etc, they have indeed said that the red medicine will do everything that the white medicine will do and more, but they are all incorrect, this is why, if you would make a medicine for man, out of the stone for metals, it must be very well diluted, and very highly filtered, to remove all of the white salt, (luna), because the medicine for man, and the medicine for metals are two different things, and rightly so.

LET ME REITERATE ONCE AGAIN AT THIS JUNCTURE, JUST A BRIEF INTERMISSION HERE, DO NOT WORK WITH ANY TOXIC SUBSTANCES, AND AVOID THE ADVICE OF THOSE WHO WOULD LEAD YOU ASTRAY BY ADVISING YOU TO EXPERIMENT WITH SUCH POISONS! USE SOME COMMON SENSE, WHEN YOU ARE COOKING AND PREPARING FOOD OR DRINK TO BE CONSUMED, ARE THERE NOT CERTAIN THINGS AND OR SUBSTANCES THAT YOU WOULD AVOID?, IT REALLY IS THE SAME IN ALCHEMY, IF YOU WOULD NOT EAT OR DRINK ANY OF THE ITEMS, THEN THEY SIMPLY DO NOT BELONG IN YOUR WORK, BECAUSE THERE IS NOTHING FOUL OR TOXIC HERE, ONLY A MEDICINE FOR MAN, AND A MEDICINE FOR METALS.

It has been written, that the finding of the stone, is in fact not the end of the work, but only the beginning, for only now has one thrown open the gates of nature, and opened up the possibility of a much deeper and profound study of her ancient mysteries. The finding of the stone is not the end of the great work of the holy science, but I say to you that this is only the beginning, for only now, can one even just begin to experiment.

TRANSMUTATION INSTRUCTIONS.

We will begin to delve here, into a very rare aspect of this art, the actual transmutation instructions themselves, which are a whole separate study above the finding of the stone. And because of this, another book will have to be written, as I delve deeper into this mystery, but for now, we will cover what we have learned thus far, which begins here.

For this experiment the old ones of previous centuries generally used cast iron square nails, which are still available today in concrete work, but in our modern times I myself would prefer to use a piece of clean copper wire that has been carefully stripped of its insulation.

Take the aqua vitae, our water of life, and take the white stone itself, as two separate substances which have not been subjected to conjunction, triturate or otherwise render the sulfur principle, our fire, the white salt, to powder, now coat the metal with the liquid, and then sprinkle it with the white salt, so that it sticks or adheres to the metal, and now bake the metal in a high heat furnace until it glows red hot, and then allow it to cool.

This is how the old ones made silver. The process for gold is different, it is more complex, and it seems that it will most likely become cause for the writing of another book above this one. Because alchemy can be a very difficult and confusing art to figure out, we will take it slow, one step at a time, unraveling this hermetic mystery. But we will in fact get there, that much is for sure.

The black stage, the ravens head, signifies distillation. This is the point in time when the matter turns black. Contrary to popular belief, this occurs in the retort, not in the flask.

Yes some of my books appear to be difficult, and some of them are short, but in fact, they are very simple, just as simplicity is the seal of truth, and this book is not only longer, but much more simpler than some of the others in this series, not to mention the fact that these books are much

much more simpler than some of the other books which are available on the market today, especially those written by false adepts who as of yet, still have not found the philosophical stone.

In the realm of the philosophers stone, there are some who believe that there are two stones, the white stone, and the red stone. The truth of the matter is that there are three stones, the first stone is the medicine for man, the second stone is the white stone for metals, the third is the citrine stone for metals. The medicine for man and the medicine for metals are two different things, and rightly so, because if you were to fix a rubber tire, you would use a rubber patch, not a dissimilar substance. The medicine for man however, does not come from man, man comes from it, we were created from this substance, and that is why it is medicinal to us. The medicine for man is a clear liquid, if it is subjected to gentle coction (cooking), it undergoes color changes, it becomes white, then yellow, and finally as red as the most brilliant fiery ruby, and it will coagulate into a stone, which can be multiplied forever into more medicine. One grain of this stone, as in the size of a grain of salt, is dissolved in liquid, (about one ounce of liquid), and this is aurum solis. This is how someone that you may recall in history turned the water into wine. The wise ancient masters only consumed a few drops of this liquid twice per year on the vernal equinoxes for a long and healthful life.

The effects of the medicine are everything that the sages said it was.
Hermes trismegistus, sometimes consumed it as the clear liquid, in which he only consumed a few drops. He also used to bake it in little cone shaped molds so that it would coagulate into a stone, and then it would simply be a little cone, similar to baking a loaf of bread. These cones were then served to the Egyptian pharaohs and high priests, these little cones were considered to be the food of the gods, (which they are), going from the drops to the small cones indicates that over time, one builds a higher level of tolerance and can gradually accept a higher dosage of the royal medicine.
Some of the effects include a strengthening of the limbs, and of the heart, as well as improvement of the vision and hearing.

THE FLAMEL PATH.

Nicholas Flamel learned from the master directory itself, the book of Abraham Eleazer.

Mr Flamel followed the dry path. This was his method. However he was also familiar with the wet path and this is how it was done in his time. I have sufficiently described my work in my series of booklets so you should be able to follow along, and my own method which I mainly discuss in my works is the wet path, with that being said, at the portion of the work where you have separated and purified the three substances, the crystal clear liquid, the white salt and the red powder, the method from here of the old ones, was that they would put the red powder, (our salamander which endures the fire and our incombustible sulfur), back into the retort, they would then return the volatile spirit to the earth and perform the flight of ten eagles, at this juncture the substance will come over the helm as a blood red oil by drops, and will have a very sweet smell, now with this red oil and the white salt, we have the red man and the white wife, we can then proceed to the conjunction, and once this cycle is completed we can repeat the same exact process which is now known as imbibition/multiplication.

Flamel was proficient in both the wet and dry paths, some people say that he preferred one over the other, which is debatable, I have my opinion, but when opinions are shared arguments soon develop and there is no need for any of that especially over such trivial issues.

Mr Paracelsus was a very arrogant and pompous man, he was very highly intelligent, the ingestion of the medicine no doubt was somewhat responsible for this as it caused him to become one of the illuminated ones, but apparently he had forgotten where he came from because he did not discover this art on his own, it was taught to him by a master alchemist and if it hadn't have been for that, he would have been a nobody.

In any case, this art had been taught to him from a teacher, after that, he was an alchemist, and alchemists love to experiment, which caused him to discover the short wet path, which greatly shortened the work, he simply discovered that since the alchemists had to perform the flight of ten eagles when they confected their menstruum, they had mistakenly assumed that it would need to be repeated upon the aqua vitae, this however is completely unnecessary, as the gentle warmth of coction will cause all of the changes in the matter, so Paracelsus eliminated the flight of ten eagles from flamels portion of the process which is known as separation and purification of the elements, which is also sometimes called, the true separation of the bodies, which occurs directly after the black stage, (the ravens head).

So instead of performing the long and tedious task of the ten distillations, Paracelsus simply skipped straight ahead to conjunction.

Mr Hohenhiems multiplications took twenty days apiece, and this is what he did, he took the white salt, and imbibed it with just as much liquid as it could absorb and no more, (make a slurry), he then sealed the flask and baked it in his athanor for twenty days until it had congealed into a stone, the multiplications must be repeated until suddenly the white salt undergoes a drastic and notable change, it will suddenly melt and flow like wax with no fumes, at this time the two substances have not only become one, but they together, have learned to endure the fire and can never again be separated from one another, at this time, the medicine now has ingress, which is to say that it has power over the metals to tinge them. This process takes three turns of the alchemical wheel to reach this point in time, so with Paracelsus method of imbibition, you can get there in sixty days. Flamel would have still been performing his flight of ten eagles at this time.

Many of the sages were satisfied at this juncture, the citrine stone of the third order. Mr Flamel however preferred one additional multiplication and brought his work to the fourth order, which can clearly be seen in his writings. Flamels stone of the fourth order, to give you an idea of its power, one ounce of this stone, was cast upon 100,000 ounces of inferior metal. According to his writings, he only made projection three times, and it is not hard to see why he didn't need any additional projection, Flamel liked to go big, and back then they had not yet invented dump trucks for transportation.

The stone is very fusible, this is why it penetrates metal and becomes one with it, the more you multiply this substance, the more fusible it becomes, and this is why the tinging power is increased with each multiplication, also because the substance which is used as the multiplier, also is the colorant, since the body of the stone itself, it simply just a white salt.

Because the stone becomes more fusible with each turn of the alchemical wheel, it can only be multiplied to the seventh power, after this no glass can hold it because it will fuse itself with the glass, and then the delicious cake which you have so tediously baked, is ruined, there would never be any need to continue further anyway, the stone of the fourth order, is already plenty, you know how it is in the drive through when you supersize your order, it really is already more than we need.

I am going to create a dvd of the magnum opus at some point in the future. It will be like a documentary, however it will be a silent film, I will perform the entire magnum opus, I will take a picture of each step of the work from start to finish, I will then arrange the pictures into a video slideshow, each picture will show for about eight seconds and there will be captions, the movie will be a slide show of words and pictures. There may or may not be back ground music involved, it depends upon if suitable music is chosen and permission granted by applicable copyright holders.

The video will show the processes and in the correct order, it will also show the proper vessels, it will not however list the prima materia or the menstruum, but you should already know these things quite well by now if you studied all of my booklets.

Before we close this installment in my series of booklets upon the great work, let us briefly touch upon the great medicine, Aurum Solis. There is much confusion in the world of alchemy, only when you actually find the philosophers stone can you begin to separate fact from fiction, so let us clear up a certain issue right here and now as regards the medicine for man.

Now some persons have said that the white stone is a great medicine, that remains to be seen, and I am beginning to have my doubts about that, in the hermetical wedding we have two substances, sol and luna, our philosophical gold and our living silver, luna is a volatile white salt, it is in fact the white stone itself, sol, which is our living gold, is an abbreviation for aurum solis, the truth of the matter is that the two substances sol and luna are a liquid and a salt, unbeknownst to many, the liquid is sol, and the white substance is luna, this is why in old pictures of the green lion devouring the sun, we see the moon going down in the water, this signifies conjunction, which occurs after the green lion devours the sun, which in that particular instance, the sun represents fire, as in heat, the green lion represents our secret prima materia, the moon and the water are sol and luna, the moon in the water represents the hermetical wedding, so sol, puts the sol in aurum solis, not luna the white stone, au means gold, so whether luna itself which is the white stone has any medicinal benefit remains to be seen, and the water alone is the elixir of life, it is also the multiplier which is used to perform the imbibitions upon the white stone in order to bring it to the citrine. When the ancients made medicine from the citrine stone itself, they took one grain of this stone, as in the size of a grain of salt, they dissolved it in liquid until it was the color of liquid gold, then they let it set in a sealed container, for any length of time, the longer the better, but the average was from 4 hours to four days, at which time they filtered the liquid, it was allowed to set until a whitish ring formed in the glass, this is the sign that it is ready to be filtered, the goal was to filter away this whitish ring, which is luna, the white salt. So a separation is occurring here, when they made the medicine for man from the completed citrine stone for metals, they removed luna first. Luna is not part of aurum solis, which does not mean for certain that luna itself is not a medicine for man, it simply means that there is good reason for doubt, so proceed very cautiously, and remember the advice of the ancients, try your medicinal tinctures upon plants first to see the effect, better to sacrifice a plant than yourself. So proceed cautiously. Hermes himself took sol before conjunction and simply used the white drops. There are three stones in the realm of the philosophers stone, we see references in the writings to the white stone, the red stone, the citrine stone, the citrine stone is more of a yellow/orange color. The white stone we have all pretty much heard enough about that to have a decent understanding of its properties, the celestial ruby itself, the dark red stone, is the medicine for man, aurum solis itself, without conjunction with luna, coagulated into a stone. A grain of this can be dissolved into a medicine, which is known to us as aurum solis. So these are two of the stones, the white stone and the celestial ruby, the third stone is the citrine stone, this is the second medicine for metals after the white stone that we already mentioned. The citrine stone is the result of the conjunction of sol and luna, the liquid aurum solis, in its original form, without having been coagulated into a stone, is used to imbibe the white stone, and then it is sealed in a round bottom flask and coagulated into the citrine stone using the Egyptian fire. This process of the imbibition/coagulation is repeated a minimum of three times, until suddenly the white stone changes, it will now melt and flow without evaporating in the heat, at this point the philosophical child is born, which can endure the fire. Further multiplications can be performed, but do not go past the seventh or the stone will fuse with its container and be lost.

This is the eight booklet in this series on the magnum opus.

I hope that you have enjoyed it.

Other books that I have written.

Alchemy and the green lion.
Alchemy and the golden water.
Alchemy and the peacocks tail.
Alchemy and the ravens head the secret of the red mercury.
Alchemy and the golden process.
Alchemy and the tincture of gold.
Alchemy survival guide.
Sol and Luna the hermetical wedding.
Casino survival guide breaking the bank.
Karate secrets revealed knowledge of the masters.
Grandmas delicious recipes.
The secret recipe book kitchen tool box.
Booze survival guide.
Wilderness survival tips.
How to make money.
Trophy wife.
Chinese takeout recipes.

About the author.

I enjoy cooking at home from scratch, I like writing and have been interested in becoming an author for many years, thanks to amazon.com my dream has become a reality. I enjoy the company of my rottweilers Belle, Rico, Sheba and Wu Tang. I also enjoy the hermetic arts and have been studying alchemy diligently since 2008 with an emphasis on the white and red philosophers stones, the elixir of life, and the primum ens mellisa.

I was born in the winter of 1969.

DISCLAIMER.

This book is written for historical reference of the lost hermetical art only.

STEVEN SCHOOL

Notes.

ALCHEMY AND THE TICTURE OF GOLD.

Notes.

CPSIA information can be obtained at www.ICGtesting.com
Printed in the USA
LVOW12s0530140514

385714LV00012B/154/P